Chapter 8
Algebra: Equations and Inequalities

Houghton Mifflin Harcourt

Made in the United States
Text printed on 100%
recycled paper

Houghton Mifflin Harcourt

Printed in the U.S.A.

ISBN 978-0-544-34248-4

13 14 15 16 17   0928   22 21 20 19 18

4500713556   C D E F G

Dear Students and Families,

Welcome to **Go Math!**, Grade 6! In this exciting mathematics program, there are hands-on activities to do and real-world problems to solve. Best of all, you will write your ideas and answers right in your book. In **Go Math!**, writing and drawing on the pages helps you think deeply about what you are learning, and you will really understand math!

By the way, all of the pages in your **Go Math!** book are made using recycled paper. We wanted you to know that you can Go Green with **Go Math!**

Sincerely,

The Authors

Made in the United States
Text printed on 100% recycled paper

Authors

Juli K. Dixon, Ph.D.
Professor, Mathematics Education
University of Central Florida
Orlando, Florida

Edward B. Burger, Ph.D.
President, Southwestern University
Georgetown, Texas

Steven J. Leinwand
Principal Research Analyst
American Institutes for
    Research (AIR)
Washington, D.C.

Contributor

Rena Petrello
Professor, Mathematics
Moorpark College
Moorpark, California

Matthew R. Larson, Ph.D.
K-12 Curriculum Specialist for
    Mathematics
Lincoln Public Schools
Lincoln, Nebraska

Martha E. Sandoval-Martinez
Math Instructor
El Camino College
Torrance, California

English Language Learners Consultant

Elizabeth Jiménez
CEO, GEMAS Consulting
Professional Expert on English
    Learner Education
Bilingual Education and
    Dual Language
Pomona, California

Expressions and Equations

 Critical Area Writing, interpreting, and using expressions and equations

8 Algebra: Equations and Inequalities 419

COMMON CORE STATE STANDARDS
6.EE Expressions and Equations
Cluster B Reason about and solve one-variable equations and inequalities
6EE.B.5, 6EE.B.7, 6EE.B.8

GO DIGITAL

Go online! Your math lessons are interactive. Use *i*Tools, Animated Math Models, the Multimedia *e*Glossary, and more.

Chapter 8 Overview

In this chapter, you will explore and discover answers to the following **Essential Questions**:

- How can you use equations and inequalities to represent situations and solve problems?
- How can you use Properties of Equality to solve equations?
- How do inequalities differ from equations?
- Why is it useful to describe situations by using algebra?

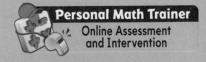

Personal Math Trainer
Online Assessment and Intervention

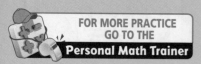

FOR MORE PRACTICE
GO TO THE
Personal Math Trainer

Practice and Homework

Lesson Check and
Spiral Review in
every lesson

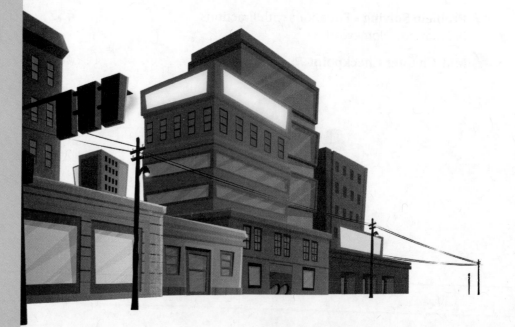

Algebra: Equations and Inequalities

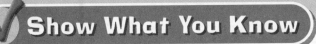

✓ Show What You Know

Personal Math Trainer
Online Assessment
and Intervention

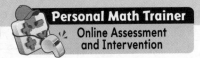

Check your understanding of important skills.

Name _____

▶ **Multiplication Properties** **Find the unknown number. Write which multiplication property you used.** (5.NBT.B.6)

1. $42 \times$ _____ $= 42$

2. $9 \times 6 =$ _____ $\times 9$

_____   _____

▶ **Evaluate Algebraic Expressions** **Evaluate the expression.** (6.EE.A.2c)

3. $4a - 2b$ for $a = 5$ and $b = 3$

4. $7x + 9y$ for $x = 7$ and $y = 1$

_____   _____

5. $8c \times d - 6$ for $c = 10$ and $d = 2$

6. $4s \div t + 10$ for $s = 9$ and $t = 3$

_____   _____

▶ **Add Fractions and Decimals** **Find the sum. Write the sum in simplest form.** (5.NF.A.1)

7. $35.68 + 17.84 =$ _____

8. $24.38 + 25.3 =$ _____

9. $\frac{3}{4} + \frac{1}{8} =$ _____

10. $\frac{2}{5} + \frac{1}{4} =$ _____

The equation $m = 19.32v$ can be used to find the mass m in grams of a pure gold coin with volume v in cubic centimeters. Carl has a coin with a mass of 37.8 grams. The coin's volume is 2.1 cubic centimeters. Could the coin be pure gold? Explain your reasoning.

Vocabulary Builder

▶ **Visualize It** •

**Use the review words to complete the tree diagram.
You may use some words more than once.**

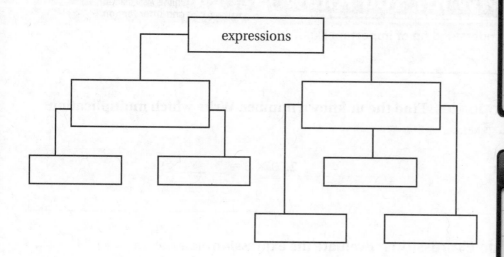

▶ **Understand Vocabulary** •

Draw a line to match the preview word with its definition.

Preview Words

1. Addition Property of •
   Equality

2. inequality •

3. inverse operations •

4. equation •

5. solution of an equation •

6. Subtraction Property •
   of Equality

Definitions

• operations that undo each
  other

• a value of a variable that makes an
  equation true

• property that states that if you add
  the same number to both sides of an
  equation, the two sides will remain equal

• a mathematical statement that compares
  two expressions by using the symbol $<$,
  $>$, $\leq$, $\geq$, or $\neq$

• property that states that if you subtract
  the same number from both sides of an
  equation, the two sides will remain equal

• a statement that two mathematical
  expressions are equal

• **Interactive Student Edition**
• **Multimedia eGlossary**

Addition Property of Equality

propiedad de suma de la igualdad

2

algebraic expression

expresión algebraica

3

Division Property of Equality

propiedad de división de la igualdad

24

equation

ecuación

27

Identity Property of Addition

propiedad de identidad de la suma

40

Identity Property of Multiplication

propiedad de identidad de la multiplicación

41

inequality

desigualdad

43

inverse operations

operaciones inversas

46

An expression that includes at least one variable

$$x + 10 \qquad 3 \times y \qquad 3 \times (a + 4)$$

variable   variable   variable

Addition Property of Equality

If you add the same number to both sides of an equation, the two sides will remain equal.

$$7 - 4 = 3$$
$$7 - 4 + 4 = 3 + 4$$
$$7 + 0 = 7$$
$$7 = 7$$

An algebraic or numerical sentence that shows that two quantities are equal

Examples:

$$8 + 12 = 20 \qquad 14 = a - 3 \qquad 2d = 14$$

Division Property of Equality

If you divide both sides of an equation by the same nonzero number, the two sides will remain equal.

$$2 \times 6 = 12$$
$$\frac{2 \times 6}{2} = \frac{12}{2}$$
$$1 \times 6 = 6$$
$$6 = 6$$

The property that states that the product of any number and 1 is that number

Examples:

$$475 \times 1 = 475 \qquad \frac{2}{3} \times 1 = \frac{2}{3} \qquad 0.7 \times 1 = 0.7$$

The property that states that when you add zero to a number, the result is that number

Examples:

$$5{,}026 + 0 = 5{,}026 \qquad \frac{2}{3} + 0 = \frac{2}{3}$$

$$1.5 + 0 = 1.5$$

Opposite operations, or operations that undo each other, such as addition and subtraction or multiplication and division

A mathematical sentence that contains the symbol $<$, $>$, $\leq$, $\geq$, or $\neq$

Examples:

$$8 < 11 \qquad 9 > {}^-4 \qquad a \leq 50 \qquad x \geq 3.2$$

Multiplication Property of Equality

propiedad de multiplicación de la igualdad

62

reciprocal

recíproco

89

solution of an equation

solución de una ecuación

94

solution of an inequality

solución de una desigualdad

95

Subtraction Property of Equality

propiedad de resta de la igualdad

98

variable

variable

105

Two numbers are reciprocals of each other if their product equals 1.

Example: $\frac{2}{3} \times \frac{3}{2} = 1$

Multiplication Property of Equality

If you multiply both sides of an equation by the same number, the two sides will remain equal.

$\frac{12}{4} = 3$

$4 \times \frac{12}{4} = 4 \times 3$

$1 \times 12 = 12$

$12 = 12$

A value that, when substituted for the variable, makes an inequality true

A value that, when substituted for the variable, makes an equation true.

Example:

$x + 3 = 5$ $x = 2$ is the solution of the equation because $2 + 3 = 5$.

A letter or symbol that stands for an unknown number or numbers

$x + 10$ $3 \times y$ $3 \times (a + 4)$

↑   ↑   ↑

variable   variable   variable

Subtraction Property of Equality

If you subtract the same number from both sides of an equation, the two sides will remain equal.

$3 + 4 = 7$

$3 + 4 - 4 = 7 - 4$

$3 + 0 = 3$

$3 = 3$

Pick It

For 3 players

Materials

- 4 sets of word cards

How to Play

1. Each player is dealt 5 cards. The remaining cards are a draw pile.

2. To take a turn, ask any player if he or she has a word or term that matches one of your word cards.

3. If the player has the word, he or she gives the word card to you, and you must define the term.

   - If you are correct, keep the card and put the matching pair in front of you. Take another turn.

   - If you are wrong, return the card. Your turn is over.

4. If the player does not have the word, he or she answers, "Pick it." Then you take a card from the draw pile.

5. If the card you draw matches one of your word cards, follow the directions for Step 3 above. If it does not, your turn is over.

6. The game is over when one player has no cards left. The player with the most pairs wins.

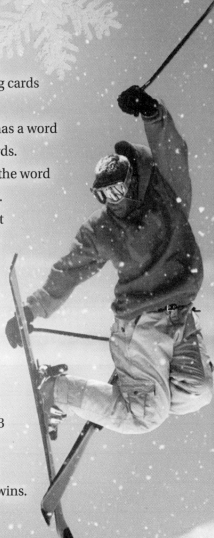

Word Box

Addition Property of Equality

algebraic expression

Division Property of Equality

equation

Identity Property of Addition

Identity Property of Multiplication

inequality

inverse operations

Multiplication Property of Equality

reciprocal

solution of an equation

solution of an inequality

Subtraction Property of Equality

variable

The Write Way

Reflect

Choose one idea. Write about it.

- Which of the following is an equation? Tell how you know.

  $8 + 20 =$ $9x + 4$ $6d = 12$ $3(5 + 2^3)$

- Explain how you can use inverse operations to solve this equation and check your solution.

  $a - 7 = 15$

- Suppose you write a math advice column and a reader needs help solving a multiplication equation. Write a letter to the reader explaining how to use the Division Property of Equality to solve the equation.

- Compare and contrast an equation and an inequality. How are they alike? How are they different?

Name _____

Solutions of Equations

Essential Question How do you determine whether a number is a solution of an equation?

Common Core — Expressions and Equations—
6.EE.B.5
MATHEMATICAL PRACTICES
MP2, MP3, MP6

An **equation** is a statement that two mathematical expressions are equal. These are examples of equations:

$$8 + 12 = 20 \qquad 14 = a - 3 \qquad 2d = 14$$

A **solution of an equation** is a value of a variable that makes an equation true.

$x + 3 = 5$ $x = 2$ is the solution of the equation because $2 + 3 = 5$.

 Unlock the Problem Real World

In the 2009–2010 season, the women's basketball team of Duke University lost 5 of their 29 games. The equation $w + 5 = 29$ can be used to find the team's number of wins w. Determine whether $w = 14$ or $w = 24$ is a solution of the equation, and tell what the solution means.

Use substitution to determine the solution.

STEP 1 Check whether $w = 14$ is a solution.

Write the equation.	$w + 5 = 29$
Substitute 14 for w.	_____ $+ 5 \stackrel{?}{=} 29$
Add.	_____ $\neq 29$

The equation is not true when $w = 14$, so $w = 14$ is not a solution.

Math Idea

The symbol $\neq$ means "is not equal to."

STEP 2 Check whether $w = 24$ is a solution.

Write the equation.	$w + 5 = 29$
Substitute 24 for w.	_____ $+ 5 \stackrel{?}{=} 29$
Add.	_____ $= 29$

The equation is true when $w = 24$, so $w = 24$ is a solution.

So, the solution of the equation $w + 5 = 29$ is $w =$ _____,

which means that the team won _____ games.

Math Talk   MATHEMATICAL PRACTICES ⑥

Explain How is an algebraic equation, such as $x + 1 = 4$, different from a numerical equation, such as $3 + 1 = 4$?

🔑 Example 1 Determine whether the given value of the variable is a solution of the equation.

Ⓐ $x - 0.7 = 4.3$; $x = 3.6$

Write the equation.

$x - 0.7 = 4.3$

Substitute the given value for the variable.

_____ $- 0.7 \overset{?}{=} 4.3$

Subtract. Write = or ≠.

_____ ◯ 4.3

The equation _____ true when $x = 3.6$, so $x = 3.6$

_____ a solution.

Ⓑ $\frac{1}{3}a = \frac{1}{4}$; $a = \frac{3}{4}$

Write the equation.

$$\frac{1}{3}a = \frac{1}{4}$$

Substitute the given value for the variable.

$\frac{1}{3} \times$ ⬚⬚ $\overset{?}{=} \frac{1}{4}$

Simplify factors and multiply. Write = or ≠.

⬚⬚ ◯ $\frac{1}{4}$

The equation _____ true when $a = \frac{3}{4}$, so $a = \frac{3}{4}$

_____ a solution.

🔑 Example 2 The sixth-grade class president serves a term of 8 months.

Janice has already served 5 months of her term as class president. The equation $m + 5 = 8$ can be used to determine the number of months m Janice has left. Use mental math to find the solution of the equation.

Think: What number plus 5 is equal to 8?   _____ plus 5 is equal to 8.

Use substitution to check whether $m = 3$ is a solution.

Write the equation.

$m + 5 = 8$

Substitute 3 for m.

_____ $+ 5 \overset{?}{=} 8$

Add. Write = or ≠.

_____ ◯ 8

So, $m =$ _____ is the solution of the equation, and

_____ months of Janice's term remain.

> **Math Talk**
>
> MATHEMATICAL PRACTICES ③
>
> **Apply** Give an example of an equation whose solution is $y = 7$. Explain how you know that the equation has this solution.

Share and Show MATH BOARD

Determine whether the given value of the variable is a solution of the equation.

1. $x + 12 = 29$; $x = 7$

_____ $+ 12 \overset{?}{=} 29$

_____ $\bigcirc$ 29

_____

2. $n - 13 = 2$; $n = 15$

_____

✓ **3.** $\frac{1}{2}c = 14$; $c = 28$

_____

✓ **4.** $m + 2.5 = 4.6$; $m = 2.9$

_____

5. $d - 8.7 = 6$; $d = 14.7$

_____

6. $k - \frac{3}{5} = \frac{1}{10}$; $k = \frac{7}{10}$

_____

Math Talk MATHEMATICAL PRACTICES ⑥

Explain why $2x - 6$ is not an equation.

On Your Own

Determine whether the given value of the variable is a solution of the equation.

7. $17.9 + v = 35.8$; $v = 17.9$

_____

8. $c + 35 = 57$; $c = 32$

_____

9. $18 = \frac{2}{3}h$; $h = 12$

_____

10. In the equation $t + 2.5 = 7$, determine whether $t = 4.5$, $t = 5$, or $t = 5.5$ is a solution of the equation.

_____

11. Antonio ran a total of 9 miles in two days. The first day he ran $5\frac{1}{4}$ miles. The equation $9 - d = 5\frac{1}{4}$ can be used to find the distance d in miles Antonio ran the second day. Determine whether $d = 4\frac{3}{4}$, $d = 4$, or $d = 3\frac{3}{4}$ is a solution of the equation, and tell what the solution means.

_____

Problem Solving • Applications

Use the table for 12–14.

12. **MATHEMATICAL PRACTICE ②** **Connect Symbols and Words** The length of a day on Saturn is 14 hours less than a day on Mars. The equation $24.7 - s = 14$ can be used to find the length in hours s of a day on Saturn. Determine whether $s = 9.3$ or $s = 10.7$ is a solution of the equation, and tell what the solution means.

_____

13. A storm on one of the planets listed in the table lasted for 60 hours, or 2.5 of the planet's days. The equation $2.5h = 60$ can be used to find the length in hours h of a day on the planet. Is the planet Earth, Mars, or Jupiter? Explain.

_____

_____

Length of Day	
Planet	**Length of Day (hours)**
Earth	24.0
Mars	24.7
Jupiter	9.9

14. **GO DEEPER** A day on Pluto is 143.4 hours longer than a day on one of the planets listed in the table. The equation $153.3 - p = 143.4$ can be used to find the length in hours p of a day on the planet. What is the length of a storm that lasts $\frac{1}{3}$ of a day on this planet?

_____

15. **THINK SMARTER** **What's the Error?** Jason said that the solution of the equation $2m = 4$ is $m = 8$. Describe Jason's error, and give the correct solution.

_____

_____

16. **THINK SMARTER** The marking period is 45 school days long. Today is the twenty-first day of the marking period. The equation $x + 21 = 45$ can be used to find the number of days x left in the marking period. Using substitution, Rachel determines

there are
20
24
26
days left in the marking period.

Solutions of Equations

COMMON CORE STANDARD—6.EE.B.5
Reason about and solve one-variable equations and inequalities.

**Determine whether the given value of the variable is
a solution of the equation.**

1. $x - 7 = 15; x = 8$

$\underline{8} - 7 \overset{?}{=} 15$

$\underline{1}$ ⊘≠ 15

_____ **not a solution** _____

2. $c + 11 = 20; c = 9$

_____

3. $\frac{1}{3}h = 6; h = 2$

_____

4. $16.1 + d = 22; d = 6.1$

_____

5. $9 = \frac{3}{4}e; e = 12$

_____

6. $15.5 - y = 7.9; y = 8.4$

_____

Problem Solving · Real World

7. Terrance needs to score 25 points to win a game. He has already scored 18 points. The equation $18 + p = 25$ can be used to find the number of points p that Terrance still needs to score. Determine whether $p = 7$ or $p = 13$ is a solution of the equation, and tell what the solution means.

_____

_____

_____

8. Madeline has used 50 sheets of a roll of paper towels, which is $\frac{5}{8}$ of the entire roll. The equation $\frac{5}{8}s = 50$ can be used to find the number of sheets s in a full roll. Determine whether $s = 32$ or $s = 80$ is a solution of the equation, and tell what the solution means.

_____

_____

_____

9. **WRITE** ▸*Math* Use mental math to find the solution to $4x = 36$. Then use substitution to check your answer.

_____

_____

1. Sheena received a gift card for $50. She has already used it to buy a lamp for $39.99. The equation $39.99 + x = 50$ can be used to find the amount x that is left on the gift card. What is the solution of the equation?

2. When Pete had a fever, his temperature was 101.4°F. After taking some medicine, his temperature was 99.2°F. The equation $101.4 - d = 99.2$ can be used to find the number of degrees d that Pete's temperature decreased. What is the solution of the equation?

Spiral Review (6.RP.A.3c, 6.EE.A.1, 6.EE.A.4, 6.EE.B.6)

3. Melanie has saved $60 so far to buy a lawn mower. This is 20% of the price of the lawn mower. What is the full price of the lawn mower that she wants to buy?

4. A team of scientists is digging for fossils. The amount of soil in cubic feet that they remove is equal to 6^3. How many cubic feet of soil do the scientists remove?

5. Andrew made p picture frames. He sold 2 of them at a craft fair. Write an expression that could be used to find the number of picture frames Andrew has left.

6. Write an expression that is equivalent to $4 + 3(5 + x)$.

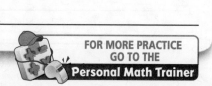

FOR MORE PRACTICE
GO TO THE
Personal Math Trainer

Write Equations

Essential Question How do you write an equation to represent a situation?

Common Core
Expressions and Equations—
6.EE.B.7
MATHEMATICAL PRACTICES
MP2, MP3, MP4, MP6

CONNECT You can use what you know about writing algebraic expressions to help you write algebraic equations.

Unlock the Problem Real World

A circus recently spent $1,650 on new trapezes. The trapezes cost $275 each. Write an equation that could be used to find the number of trapezes *t* that the circus bought.

- Circle the information that you need to write the equation.
- What expression could you use to represent the cost of *t* trapezes?

_____

Write an equation for the situation.

Think:

Cost per trapeze	times	number of trapezes	equals	total cost.
↓	↓	↓	↓	↓
_____	×	*t*	=	_____

So, an equation that could be used to find the number of

trapezes *t* is _____.

Try This! Ben is making a recipe for salsa that calls for $3\frac{1}{2}$ cups of tomatoes. He chops 4 tomatoes, which fill $2\frac{1}{4}$ cups. Write an equation that could be used to find out how many more cups *c* Ben needs.

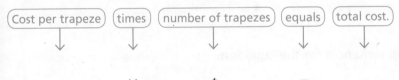

Think:

Cups filled	plus	cups needed	equals	total cups for recipe.
↓	↓	↓	↓	↓
_____	+	_____	=	_____

So, an equation that could be used to find the number of

additional cups *c* is _____.

Math Talk
MATHEMATICAL PRACTICES ④

Represent What is another equation you could use to model the problem?

🔑 Example 1 Write an equation for the word sentence.

Ⓐ Six fewer than a number is 46.33.

Think: Let *n* represent the unknown number. The phrase "fewer than" indicates

_____.

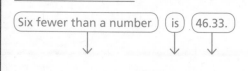

Six fewer than a number → ___ — ___ = ___
is →
46.33. →

Ⓑ Two-thirds of the cost of the sweater is $18.

Think: Let *c* represent the _____ of the sweater in dollars. The word "of"

indicates _____.

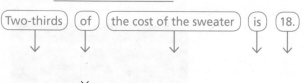

Two-thirds → ___ × ___ = ___
of →
the cost of the sweater →
is →
18. →

🔑 Example 2 Write two word sentences for the equation.

Ⓐ $a + 15 = 24$

- The _____ of *a* and 15 _____ 24.
- 15 _____ than *a* _____ 24.

Ⓑ $r \div 0.2 = 40$

- The _____ of *r* and 0.2 _____ 40.
- *r* _____ by 0.2 _____ 40.

---

1. Explain how you can rewrite the equation $n + 8 = 24$ so that it involves subtraction rather than addition.

_____

_____

2. **MATHEMATICAL PRACTICE ❸ Compare Representations** One student wrote $18 \times d = 54$ for the sentence "The product of 18 and *d* equals 54." Another student wrote $d \times 18 = 54$ for the same sentence. Are both students correct? Justify your answer.

_____

_____

Name _____

1. Write an equation for the word sentence "25 is 13 more than a number."

What operation does the phrase "more than" indicate? _____

The equation is _____ = _____ + _____.

Write an equation for the word sentence.

2. The difference of a number and 2 is $3\frac{1}{3}$.

_____

3. Ten times the number of balloons is 120.

_____

Write a word sentence for the equation.

4. $x - 0.3 = 1.7$

_____

5. $25 = \frac{1}{4}n$

_____

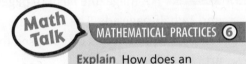

Math Talk MATHEMATICAL PRACTICES ⑥

Explain How does an equation differ from an expression?

On Your Own

Write an equation for the word sentence.

6. The quotient of a number and 20.7 is 9.

_____

7. 24 less than the number of snakes is 35.

_____

8. 75 is $18\frac{1}{2}$ more than a number.

_____

9. d degrees warmer than 50 degrees is 78 degrees.

_____

Write a word sentence for the equation.

10. $15g = 135$

_____

11. $w \div 3.3 = 0.6$

_____

Problem Solving • Applications Real World

To find out how far a car can travel on a certain amount of gas, multiply the car's fuel efficiency in miles per gallon by the gas used in gallons. Use this information and the table for 12–13.

12. Write an equation that could be used to find how many miles a hybrid SUV can travel in the city on 20 gallons of gas.

_____

13. A sedan traveled 504 miles on the highway on a full tank of gas. Write an equation that could be used to find the number of gallons the tank holds.

_____

Fuel Efficiency

Vehicle	Miles per gallon, city	Miles per gallon, highway
Hybrid SUV	36	31
Minivan	19	26
Sedan	20	28
SUV	22	26

WRITE ▸ *Math*
Show Your Work

14. **MATHEMATICAL PRACTICE ❷** **Connect Symbols to Words** Sonya was born in 1998. Carmen was born 11 years after Sonya. If you wrote an equation to find the year in which Carmen was born, what operation would you use in your equation?

_____

15. **GO DEEPER** A magazine has 110 pages. There are 23 full-page ads and 14 half-page ads. The rest of the magazine consists of articles. Write an equation that can be used to find the number of pages of articles in the magazine.

_____

16. **THINK SMARTER** **What's the Error?** Tony is traveling 560 miles to visit his cousins. He travels 313 miles the first day. He says that he can use the equation $m - 313 = 560$ to find the number of miles m he has left on his trip. Describe and correct Tony's error.

17. **THINK SMARTER** Jamie is making cookies for a bake sale. She triples the recipe in order to have enough cookies to sell. Jamie uses 12 cups of flour to make the triple batch.

Write an equation that can be used to find out how much flour f is needed for one batch of cookies.

Write Equations

Common Core

COMMON CORE STANDARD—6.EE.B.7
Reason about and solve one-variable equations
and inequalities.

Write an equation for the word sentence.

1. 18 is 4.5 times a number.

$$18 = 4.5n$$

2. Eight more than the number of children is 24.

_____

3. The difference of a number and $\frac{2}{3}$ is $\frac{3}{8}$.

_____

4. A number divided by 0.5 is 29.

_____

Write a word sentence for the equation.

5. $x - 14 = 52$

_____

6. $2.3m = 0.46$

_____

7. $25 = k \div 5$

_____

8. $4\frac{1}{3} + q = 5\frac{1}{6}$

_____

Problem Solving Real World

9. An ostrich egg weighs 2.9 pounds. The difference between the weight of this egg and the weight of an emu egg is 1.6 pounds. Write an equation that could be used to find the weight w, in pounds, of the emu egg.

_____

10. In one week, the number of bowls a potter made was 6 times the number of plates. He made 90 bowls during the week. Write an equation that could be used to find the number of plates p that the potter made.

_____

11. **WRITE** ▸*Math* When writing a word sentence as an equation, explain when to use a variable.

_____

_____

Lesson Check (6.EE.B.7)

1. Three friends are sharing the cost of a bucket of popcorn. The total cost of the popcorn is $5.70. Write an equation that could be used to find the amount a in dollars that each friend should pay.

2. Salimah had 42 photos on her phone. After she deleted some of them, she had 23 photos left. What equation could be used to find the number of photos p that Salimah deleted?

Spiral Review (6.RP.A.3d, 6.EE.A.1, 6.EE.A.3, 6.EE.B.5)

3. A rope is 72 feet long. What is the length of the rope in yards?

4. Julia evaluated the expression $3^3 + 20 \div 2^2$. What value should she get as her answer?

5. The sides of a triangle have lengths s, $s + 4$, and $3s$. Write an expression in simplest form that represents the perimeter of the triangle.

6. Gary knows that $p = 2\frac{1}{2}$ is a solution to one of the following equations. Which one has $p = 2\frac{1}{2}$ as its solution?

$$p + 2\frac{1}{2} = 5 \qquad p - 2\frac{1}{2} = 5$$

$$2 + p = 2\frac{1}{2} \qquad 4 - p = 2\frac{1}{2}$$

FOR MORE PRACTICE GO TO THE
Personal Math Trainer

Name _____

Model and Solve Addition Equations

Essential Question How can you use models to solve addition equations?

Common Core Expressions and Equations—
6.EE.B.7
MATHEMATICAL PRACTICES
MP4, MP5, MP6

You can use algebra tiles to help you find solutions of equations.

Algebra Tiles

x tile   1 tile

Investigate

Materials ■ MathBoard, algebra tiles

Thomas has $2. He wants to buy a poster that costs $7. Model and solve the equation $x + 2 = 7$ to find the amount x in dollars that Thomas needs to save in order to buy the poster.

A. Draw 2 rectangles on your MathBoard to represent the two sides of the equation.

B. Use algebra tiles to model the equation. Model $x + 2$ in the left rectangle, and model 7 in the right rectangle.

- What type of tiles and number of tiles did you use to model $x + 2$?

_____

C. To solve the equation, get the x tile by itself on one side. If you remove a tile from one side, you can keep the two sides equal by removing the same type of tile from the other side.

- How many 1 tiles do you need to remove from each side to get the x tile by itself on the left side? _____

- When the x tile is by itself on the left side, how many 1 tiles are on the right side? _____

D. Write the solution of the equation: $x =$ _____.

So, Thomas needs to save $ _____ in order to buy the poster.

Math Talk
MATHEMATICAL PRACTICES ④
Model What operation did you model when you removed the tiles?

Draw Conclusions

1. **MATHEMATICAL PRACTICE ⑤** **Use Appropriate Tools** Describe how you could use your model to check your solution.

    _____

    _____

    _____

2. Tell how you could use algebra tiles to model the equation $x + 4 = 8$.

    _____

    _____

3. *THINK SMARTER* What would you do to solve the equation $x + 9 = 12$ without using a model?

    _____

Make Connections

You can solve an equation by drawing a model to represent algebra tiles.

Let a rectangle represent the variable. Let a small square represent 1.

Solve the equation $x + 3 = 7$.

STEP 1

Draw a model of the equation.

STEP 2

Get the variable by itself on one side of the model by doing the same thing to both sides.

Cross out _____ squares on the left side and

_____ squares on the right side.

STEP 3

Draw a model of the solution.

There is 1 rectangle on the left side. There are

_____ squares on the right side.

So, the solution of the equation $x + 3 = 7$ is $x =$ _____.

434

Name _____

Share and Show MATH BOARD

Model and solve the equation by using algebra tiles or *i*Tools.

1. $x + 5 = 7$ _____

2. $8 = x + 1$ _____

3. $x + 2 = 5$ _____

4. $x + 6 = 8$ _____

5. $5 + x = 9$ _____

6. $5 = 4 + x$ _____

Solve the equation by drawing a model.

7. $x + 1 = 5$ _____

8. $3 + x = 4$ _____

9. $6 = x + 4$ _____

10. $8 = 2 + x$ _____

11. **MATHEMATICAL PRACTICE 6** **Describe a Method** Describe how you would draw a model to solve the equation $x + 5 = 10$.

_____

_____

_____

_____

© Houghton Mifflin Harcourt Publishing Company

Problem Solving • Applications (Real World)

12. **MATHEMATICAL PRACTICE 4** **Interpret a Result** The table shows how long several animals have lived at a zoo. The giraffe has lived at the zoo 4 years longer than the mountain lion. The equation $5 = 4 + y$ can be used to find the number of years y the mountain lion has lived at the zoo. Solve the equation. Then tell what the solution means.

_____

_____

Zoo Animals	
Animal	**Time at zoo (years)**
Giraffe	5
Hippopotamus	6
Kangaroo	2
Zebra	9

13. **GO DEEPER** Carlos walked 2 miles on Monday and 5 miles on Saturday. The number of miles he walked on those two days is 3 miles more than the number of miles he walked on Friday. Write and solve an addition equation to find the number of miles Carlos walked on Friday.

_____

14. **THINK SMARTER** **Sense or Nonsense?** Gabriela is solving the equation $x + 1 = 6$. She says that the solution must be less than 6. Is Gabriela's statement sense or nonsense? Explain.

_____

Personal Math Trainer

15. **THINK SMARTER +** The Hawks beat the Tigers by 5 points in a football game. The Hawks scored a total of 12 points.

Use numbers and words to explain how this model can be used to solve the equation $x + 5 = 12$.

⬜⬜⬜⬜⬜⬜ = ⬜⬜⬜⬜⬜⬜
⬜⬜ ⬜⬜⬜⬜⬜

_____

_____

_____

Model and Solve Addition Equations

COMMON CORE STANDARD—6.EE.B.7
Reason about and solve one-variable equations and inequalities.

Model and solve the equation by using algebra tiles.

1. $x + 6 = 9$

2. $8 + x = 10$

3. $9 = x + 1$

$x = 3$

_____ _____ _____

Solve the equation by drawing a model.

4. $x + 4 = 7$

5. $x + 6 = 10$

_____ _____

Problem Solving *Real World*

6. The temperature at 10:00 was 10°F. This is 3°F warmer than the temperature at 8:00. Model and solve the equation $x + 3 = 10$ to find the temperature x in degrees Fahrenheit at 8:00.

7. Jaspar has 7 more checkers left than Karen does. Jaspar has 9 checkers left. Write and solve an addition equation to find out how many checkers Karen has left.

_____ _____

8. **WRITE** ▸*Math* Explain how to use a drawing to solve an addition equation such as $x + 8 = 40$.

_____

_____

Lesson Check (6.EE.B.7)

1. What is the solution of the equation that is modeled by the algebra tiles?

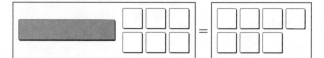

2. Alice has played soccer for 8 more years than Sanjay has. Alice has played for 12 years. The equation $y + 8 = 12$ can be used to find the number of years y Sanjay has played. How long has Sanjay played soccer?

Spiral Review (6.RP.A.3d, 6.EE.A.2a, 6.EE.A.3, 6.EE.B.7)

3. A car's gas tank has a capacity of 16 gallons. What is the capacity of the tank in pints?

4. Craig scored p points in a game. Marla scored twice as many points as Craig but 5 fewer than Nelson scored. How many points did Nelson score?

5. Simplify $3x + 2(4y + x)$.

6. The Empire State Building in New York City is 443.2 meters tall. This is 119.2 meters taller than the Eiffel Tower in Paris. Write an equation that can be used to find the height h in meters of the Eiffel Tower.

FOR MORE PRACTICE
GO TO THE
Personal Math Trainer

Name _____

Solve Addition and Subtraction Equations

Common Core Expressions and Equations—
6.EE.B.7
MATHEMATICAL PRACTICES
MP2, MP6, MP8

Essential Question How do you solve addition and subtraction equations?

CONNECT To solve an equation, you must get the variable on one side of the equal sign by itself. You have solved equations by using models. You can also solve equations by using Properties of Equality.

Subtraction Property of Equality	
If you subtract the same number from both sides of an equation, the two sides will remain equal.	$3 + 4 = 7$ $3 + 4 - 4 = 7 - 4$ $3 + 0 = 3$ $3 = 3$

Unlock the Problem

The longest distance jumped on a pogo stick is 23 miles. Emilio has jumped 5 miles on a pogo stick. The equation $d + 5 = 23$ can be used to find the remaining distance d in miles he must jump to match the record. Solve the equation, and explain what the solution means.

Solve the addition equation.

To get d by itself, you must undo the addition by 5. Operations that undo each other are called **inverse operations**. Subtracting 5 is the inverse operation of adding 5.

Write the equation. $d + 5 = 23$

Use the Subtraction Property of Equality. $d + 5 - 5 = 23 - $ _____

Subtract. $d + 0 = $ _____

Use the Identity Property of Addition.   _____ $= 18$

Check the solution.

Write the equation. $d + 5 = 23$

Substitute _____ for d.   _____ $+ 5 = 23$

The solution checks.   _____ $= 23$

So, the solution means that Emilio must jump _____ more miles.

Math Talk

MATHEMATICAL PRACTICES ⑧

Generalize How do you know what number to subtract from both sides of the equation?

When you solve an equation that involves subtraction, you can use addition to get the variable by itself on one side of the equal sign.

Addition Property of Equality	
If you add the same number to both sides of an equation, the two sides will remain equal.	$7 - 4 = 3$ $7 - 4 + 4 = 3 + 4$ $7 + 0 = 7$ $7 = 7$

🔑 Example

While cooking dinner, Carla pours $\frac{5}{8}$ cup of milk from a carton. This leaves $\frac{7}{8}$ cup of milk in the carton. Write and solve an equation to find how much milk was in the carton when Carla started cooking.

STEP 1 Write an equation.

Let a represent the amount of milk in cups in the carton when Carla started cooking.

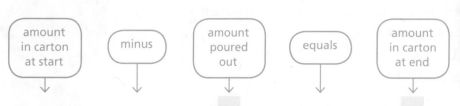

amount in carton at start	minus	amount poured out	equals	amount in carton at end
a	$-$		$=$	

STEP 2 Solve the equation.

Think: $\frac{5}{8}$ is subtracted from a, so add $\frac{5}{8}$ to both sides to undo the subtraction.

Write the equation.

$$a - \frac{5}{8} = \frac{7}{8}$$

Use the Addition Property of Equality.

$$a - \frac{5}{8} + \underline{} = \frac{7}{8} + \underline{}$$

Add.

$$a = \underline{}$$

Write the fraction greater than 1 as a mixed number, and simplify.

$$a = \underline{}$$

So, there were _____ cups of milk in the carton when Carla started cooking.

MATHEMATICAL PRACTICES ⑥

Explain How can you check the solution of the equation?

Name _____

1. Solve the equation $n + 35 = 80$.

$$n + 35 = 80$$

$$n + 35 - 35 = 80 - \underline{\hspace{1cm}} \qquad \text{Use the} \underline{\hspace{3cm}} \text{Property of Equality.}$$

$$n = \underline{\hspace{1cm}} \qquad \text{Subtract.}$$

Solve the equation, and check the solution.

2. $16 + x = 42$

3. $y + 6.2 = 9.1$

4. $m + \dfrac{3}{10} = \dfrac{7}{10}$

5. $z - \dfrac{1}{3} = 1\dfrac{2}{3}$

6. $12 = x - 24$

7. $25.3 = w - 14.9$

Math Talk

MATHEMATICAL PRACTICES ⑧

Generalize What can you do to get the variable by itself on one side of a subtraction equation?

On Your Own

Practice: Copy and Solve Solve the equation, and check the solution.

8. $y - \dfrac{3}{4} = \dfrac{1}{2}$

9. $75 = n + 12$

10. $m + 16.8 = 40$

11. $w - 36 = 56$

12. $8\dfrac{2}{5} = d + 2\dfrac{2}{5}$

13. $8.7 = r - 1.4$

14. The temperature dropped 8 degrees between 6:00 P.M. and midnight. The temperature at midnight was 26°F. Write and solve an equation to find the temperature at 6:00 P.M.

_____

15. MATHEMATICAL PRACTICE ② **Reason Abstractly** Write an addition equation that has the solution $x = 9$.

_____

🔑 Unlock the Problem

16. **GO DEEPER** In July, Kimberly made two deposits into her bank account. She made no withdrawals. At the end of July, her account balance was $120.62. Write and solve an equation to find Kimberly's balance at the beginning of July.

$	Bank Statement: Kimberly Gilson	
	Deposits	
July 12		$45.50
July 25		$43.24
	Withdrawals	
None		

a. What do you need to find?

_____

_____

b. What information do you need from the bank statement?

_____

_____

c. Write an equation you can use to solve the problem. Explain what the variable represents.

_____

_____

_____

d. Solve the equation. Show your work and describe each step.

e. Write Kimberly's balance at the beginning of July.

_____

17. **THINK SMARTER** If $x + 6 = 35$, what is the value of $x + 4$? Explain how to find the value without solving the equation.

_____

_____

_____

18. **THINK SMARTER** Select the equations that have the solution $n = 23$. Mark all that apply.

Ⓐ $16 + n = 39$

Ⓑ $n - 4 = 19$

Ⓒ $25 = n - 2$

Ⓓ $12 = n - 11$

Solve Addition and Subtraction Equations

Common Core COMMON CORE STANDARD—6.EE.B.7
Reason about and solve one-variable equations and inequalities.

Solve the equation, and check the solution.

1. $y - 14 = 23$

$y - 14 + 14 = 23 + 14$
$y = 37$

2. $x + 3 = 15$

3. $n + \frac{2}{5} = \frac{4}{5}$

4. $16 = m - 14$

5. $w - 13.7 = 22.8$

6. $s + 55 = 55$

7. $23 = x - 12$

8. $p - 14 = 14$

9. $m - 2\frac{3}{4} = 6\frac{1}{2}$

Problem Solving Real World

10. A recipe calls for $5\frac{1}{2}$ cups of flour. Lorenzo only has $3\frac{3}{4}$ cups of flour. Write and solve an equation to find the additional amount of flour Lorenzo needs to make the recipe.

11. Jan used 22.5 gallons of water in the shower. This amount is 7.5 gallons less than the amount she used for washing clothes. Write and solve an equation to find the amount of water Jan used to wash clothes.

12. **WRITE** ▸*Math* Explain how to check if your solution to an equation is correct.

Lesson Check (6.EE.B.7)

1. The price tag on a shirt says $21.50. The final cost of the shirt, including sales tax, is $23.22. The equation $21.50 + t = 23.22$ can be used to find the amount of sales tax t in dollars. What is the sales tax?

2. The equation $l - 12.5 = 48.6$ can be used to find the original length l in centimeters of a wire before it was cut. What was the original length of the wire?

Spiral Review (6.RP.A.3d, 6.EE.A.2b, 6.EE.A.4, 6.EE.B.7)

3. How would you convert a mass in centigrams to a mass in milligrams?

4. In the expression $4 + 3x + 5y$, what is the coefficient of x?

5. Write an expression that is equivalent to $10c$.

6. Miranda bought a $7-movie ticket and popcorn for a total of $10. The equation $7 + x = 10$ can be used to find the cost x in dollars of the popcorn. How much did the popcorn cost?

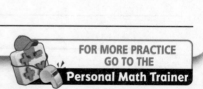

FOR MORE PRACTICE
GO TO THE
Personal Math Trainer

Model and Solve Multiplication Equations

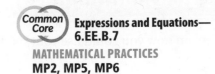

Common Core **Expressions and Equations—6.EE.B.7**

MATHEMATICAL PRACTICES
MP2, MP5, MP6

Essential Question How can you use models to solve multiplication equations?

You can use algebra tiles to model and solve equations that involve multiplication.

Algebra Tiles

x tile   1 tile

$4x$

To model an expression involving multiplication of a variable, you can use more than one x tile. For example, to model the expression $4x$, you can use four x tiles.

Investigate

Materials ■ MathBoard, algebra tiles

Tennis balls are sold in cans of 3 tennis balls each. Daniel needs 15 tennis balls for a tournament. Model and solve the equation $3x = 15$ to find the number of cans x that Daniel should buy.

A. Draw 2 rectangles on your MathBoard to represent the two sides of the equation.

B. Use algebra tiles to model the equation. Model $3x$ in the left rectangle, and model 15 in the right rectangle.

C. There are three x tiles on the left side of your model. To solve the equation by using the model, you need to find the value of one x tile. To do this, divide each side of your model into 3 equal groups.

  • When the tiles on each side have been divided into 3 equal groups, how many 1 tiles are in each group on

    the right side? _____

D. Write the solution of the equation: $x =$ _____.

So, Daniel should buy _____ cans of tennis balls.

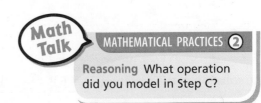

Math Talk

MATHEMATICAL PRACTICES ②

Reasoning What operation did you model in Step C?

Draw Conclusions

1. Explain how you could use your model to check your solution.

_____

_____

_____

2. **MATHEMATICAL PRACTICE ⑥** Describe how you could use algebra tiles to model the equation $6x = 12$.

_____

_____

3. **THINK SMARTER** What would you do to solve the equation $5x = 35$ without using a model?

_____

Make Connections

You can also solve multiplication equations by drawing a model to represent algebra tiles. Let a rectangle represent x. Let a square represent 1. Solve the equation $2x = 6$.

STEP 1 Draw a model of the equation.

STEP 2 Find the value of one rectangle.

Divide each side of the model into _____ equal groups.

STEP 3 Draw a model of the solution.

There is 1 rectangle on the left side. There

are _____ squares on the right side.

So, the solution of the equation $2x = 6$ is $x =$ _____.

Name _____

Model and solve the equation by using algebra tiles.

1. $4x = 16$

_____

2. $3x = 12$

_____

3. $4 = 4x$

_____

✓4. $3x = 9$

_____

5. $2x = 10$

_____

6. $15 = 5x$

_____

Solve the equation by drawing a model.

✓7. $4x = 8$ _____

8. $3x = 18$ _____

Problem Solving • Applications

9. MATHEMATICAL PRACTICE ⑤ **Communicate** Explain the steps you use to solve a multiplication equation with algebra tiles.

_____

_____

_____

_____

_____

The bar graph shows the number of countries that competed in the first four modern Olympic Games. Use the bar graph for 10–11.

10. **GO DEEPER** Naomi is doing a report about the 1900 and 1904 Olympic Games. Each page will contain information about 4 of the countries that competed each year. Write and solve an equation to find the number of pages Naomi will need.

_____

11. **THINK SMARTER** **Pose a Problem** Use the information in the bar graph to write and solve a problem involving a multiplication equation.

_____

_____

_____

_____

_____

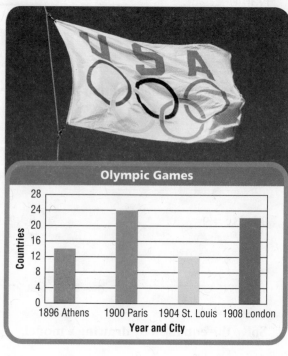

Olympic Games

Bar graph. Y-axis: Countries, marked 0, 4, 8, 12, 16, 20, 24, 28. X-axis: Year and City — 1896 Athens (14), 1900 Paris (24), 1904 St. Louis (12), 1908 London (22).

12. The equation $7s = 21$ can be used to find the number of snakes s in each cage at a zoo. Solve the equation. Then tell what the solution means.

_____

13. **THINK SMARTER** A choir is made up of 6 vocal groups. Each group has an equal number of singers. There are 18 singers in the choir. Solve the equation $6p = 18$ to find the number of singers in each group. Use a model.

Model and Solve Multiplication Equations

Common Core   **COMMON CORE STANDARD—6.EE.B.7**
Reason about and solve one-variable equations and inequalities.

Model and solve the equation by using algebra tiles.

1. $2x = 8$

2. $5x = 10$

3. $21 = 3x$

_____ $x = 4$ _____   _____   _____

Solve the equation by drawing a model.

4. $6 = 3x$

5. $4x = 12$

_____   _____

Problem Solving Real World

6. A chef used 20 eggs to make 5 omelets. Model and solve the equation $5x = 20$ to find the number of eggs x in each omelet.

7. Last month, Julio played 3 times as many video games as Scott did. Julio played 18 video games. Write and solve an equation to find the number of video games Scott played.

_____   _____

8. **WRITE** ▸*Math* Write a multiplication equation, and explain how you can solve it by using a model.

_____

_____

_____

Lesson Check (6.EE.B.7)

1. What is the solution of the equation that is modeled by the algebra tiles?

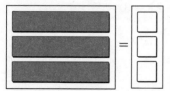

2. Carlos bought 5 tickets to a play for a total of $20. The equation $5c = 20$ can be used to find the cost c in dollars of each ticket. How much does each ticket cost?

Spiral Review (6.RP.A.3d, 6.EE.A.2c, 6.EE.B.5, 6.EE.B.7)

3. A rectangle is 12 feet wide and 96 inches long. What is the area of the rectangle?

4. Evaluate the algebraic expression $24 - x \div y$ for $x = 8$ and $y = 2$.

5. Ana bought a 15.5-pound turkey at the grocery store this month. The equation $p - 15.5 = 2.5$ can be used to find the weight p, in pounds, of the turkey she bought last month. What is the solution of the equation?

6. A pet store usually keeps 12 birds per cage, and there are 7 birds in the cage now. The equation $7 + x = 12$ can be used to find the remaining number of birds x that can be placed in the cage. What is the solution of the equation?

**FOR MORE PRACTICE
GO TO THE
Personal Math Trainer**

Name _____

Solve Multiplication and Division Equations

Essential Question How do you solve multiplication and division equations?

Common Core Expressions and Equations—
6.EE.B.7
MATHEMATICAL PRACTICES
MP1, MP7, MP8

CONNECT You can use Properties of Equality and inverse operations to solve multiplication and division equations.

Division Property of Equality

If you divide both sides of an equation by the same nonzero number, the two sides will remain equal.

$2 \times 6 = 12$

$\dfrac{2 \times 6}{2} = \dfrac{12}{2}$

$1 \times 6 = 6$

$6 = 6$

Unlock the Problem

Mei ran 14 laps around a track for a total of 4,200 meters. The equation $14d = 4,200$ can be used to find the distance d in meters she ran in each lap. Solve the equation, and explain what the solution means.

● What operation is indicated by $14d$?

_____

 Solve a multiplication equation.

To get d by itself, you must undo the multiplication by 14. Dividing by 14 is the inverse operation of multiplying by 14.

Write the equation.	$14d = 4,200$
Use the Division Property of Equality.	$\dfrac{14d}{} = \dfrac{4,200}{}$
Divide.	$1 \times d = \underline{}$
Use the Identity Property of Multiplication.	$\underline{} = 300$

Check the solution.

Write the equation. $\qquad\qquad 14d = 4,200$

Substitute _____ for d. $\qquad 14 \times$ _____ $= 4,200$

The solution checks. $\qquad\qquad$ _____ $= 4,200$

So, the solution means that Mei ran _____ meters in each lap.

Math Talk

MATHEMATICAL PRACTICES ⑧

Use Repeated Reasoning
How do you know what number to divide both sides of the equation by?

🔒 Example 1 Solve the equation $\frac{2}{3}n = \frac{1}{4}$.

Think: n is multiplied by $\frac{2}{3}$, so divide both sides by $\frac{2}{3}$ to undo the division.

Write the equation.

$$\frac{2}{3}n = \frac{1}{4}$$

Use the _____ Property of Equality.

$$\frac{2}{3}n \div \frac{2}{3} = \frac{1}{4} \div \frac{\boxed{}}{\boxed{}}$$

To divide by $\frac{2}{3}$, multiply by its reciprocal.

$$\frac{2}{3}n \times \frac{3}{2} = \frac{1}{4} \times \frac{\cdot}{\underline{}}$$

Multiply.

$$n = \frac{\boxed{}}{\boxed{}}$$

Multiplication Property of Equality

If you multiply both sides of an equation by the same number, the two sides will remain equal.

$$\frac{12}{4} = 3$$
$$4 \times \frac{12}{4} = 4 \times 3$$
$$1 \times 12 = 12$$
$$12 = 12$$

🔒 Example 2

A biologist divides a water sample equally among 8 test tubes. Each test tube contains 24.5 milliliters of water. Write and solve an equation to find the volume of the water sample.

STEP 1 Write an equation. Let v represent the volume in milliliters.

Think: The volume divided by 8 equals the volume in each test tube.

_____

$$\frac{v}{\boxed{}} = \underline{}$$

STEP 2 Solve the equation. v is divided by 8, so multiply both sides by 8 to undo the division.

Write the equation.

$$\frac{v}{8} = 24.5$$

Use the _____ Property of Equality.

$$\underline{} \times \frac{v}{8} = \underline{} \times 24.5$$

Multiply.

$$v = \underline{}$$

So, the volume of the water sample is _____ milliliters.

Math Talk

MATHEMATICAL PRACTICES ⑦

Look for Structure How can you use the Multiplication Property of Equality to solve Example 1?

Name _____

1. Solve the equation $2.5m = 10$.

$$2.5m = 10$$

$$\frac{2.5m}{2.5} = \frac{10}{}$$

Use the _____ Property of Equality.

$$m = \underline{}$$

Divide.

Solve the equation, and check the solution.

2. $3x = 210$

3. $2.8 = 4t$

✓ **4.** $\frac{1}{3}n = 15$

_____

_____

_____

5. $\frac{1}{2}y = \frac{1}{10}$

✓ **6.** $25 = \frac{a}{5}$

7. $1.3 = \frac{c}{4}$

_____

_____

_____

Math Talk

MATHEMATICAL PRACTICES ⑧

Generalize What strategy can you use to get the variable by itself on one side of a division equation?

On Your Own

Practice: Copy and Solve Solve the equation, and check the solution.

8. $150 = 6m$

9. $14.7 = \frac{b}{7}$

10. $\frac{1}{4} = \frac{3}{5}s$

11. GO DEEPER There are 100 calories in 8 fluid ounces of orange juice and 140 calories in 8 fluid ounces of pineapple juice. Tia mixed 4 fluid ounces of each juice. Write and solve an equation to find the number of calories in each fluid ounce of Tia's juice mixture.

_____

12. THINK SMARTER Write a division equation that has the solution $x = 16$.

_____

Problem Solving · Applications (Real World)

What's the Error?

13. **THINK SMARTER** Melinda has a block of clay that weighs 14.4 ounces. She divides the clay into 6 equal pieces. To find the weight w in ounces of each piece, Melinda solved the equation $6w = 14.4$.

Look at how Melinda solved the equation. Find her error.

Correct the error. Solve the equation, and explain your steps.

This is how Melinda solved the equation:

$$6w = 14.4$$

$$\frac{6w}{6} = 6 \times 14.4$$

$$w = 86.4$$

Melinda concludes that each piece of clay weighs 86.4 ounces.

So, $w =$ _____.

This means each piece of clay weighs _____.

- **MATHEMATICAL PRACTICE ①** **Describe** the error that Melinda made.

_____

_____

14. **THINK SMARTER** For numbers 14a–14d, choose Yes or No to indicate whether the equation has the solution $x = 15$.

14a. $15x = 30$ ○ Yes ○ No

14b. $4x = 60$ ○ Yes ○ No

14c. $\frac{x}{5} = 3$ ○ Yes ○ No

14d. $\frac{x}{3} = 5$ ○ Yes ○ No

Solve Multiplication and Division Equations

Common Core
COMMON CORE STANDARD—6.EE.B.7
Reason about and solve one-variable equations and inequalities.

Solve the equation, and check the solution.

1. $8p = 96$

$$\frac{8p}{8} = \frac{96}{8} \quad p = 12$$

2. $\frac{z}{16} = 8$

3. $3.5x = 14.7$

4. $32 = 3.2c$

5. $\frac{2}{5}w = 40$

6. $\frac{a}{14} = 6.8$

7. $1.6x = 1.6$

8. $23.8 = 3.5b$

9. $\frac{3}{5} = \frac{2}{3}t$

 Problem Solving *Real World*

10. Anne runs 6 laps on a track. She runs a total of 1 mile, or 5,280 feet. Write and solve an equation to find the distance, in feet, that she runs in each lap.

11. In a serving of 8 fluid ounces of pomegranate juice, there are 32.8 grams of carbohydrates. Write and solve an equation to find the amount of carbohydrates in each fluid ounce of the juice.

12. **WRITE** ▸*Math* Write and solve a word problem that can be solved by solving a multiplication equation.

Lesson Check (6.EE.B.7)

1. Estella buys 1.8 pounds of walnuts for a total of $5.04. She solves the equation $1.8p = 5.04$ to find the price p in dollars of one pound of walnuts. What does one pound of walnuts cost?

2. Gabriel wants to solve the equation $\frac{5}{8}m = 25$. What step should he do to get m by itself on one side of the equation?

Spiral Review (6.RP.A.3d, 6.EE.B.6, 6.EE.B.7)

3. At top speed, a coyote can run at a speed of 44 miles per hour. If a coyote could maintain its top speed, how far could it run in 15 minutes?

4. An online store sells DVDs for $10 each. The shipping charge for an entire order is $5.50. Frank orders d DVDs. Write an expression that represents the total cost of Frank's DVDs.

5. A ring costs $27 more than a pair of earrings. The ring costs $90. Write an equation that can be used to find the cost c in dollars of the earrings.

6. The equation $3s = 21$ can be used to find the number of students s in each van on a field trip. How many students are in each van?

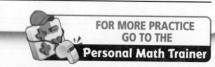

FOR MORE PRACTICE
GO TO THE
Personal Math Trainer

Name _____

Problem Solving • Equations with Fractions

Essential Question How can you use the strategy *solve a simpler problem* to solve equations involving fractions?

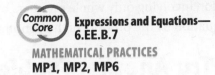

Common Core **Expressions and Equations—6.EE.B.7**
MATHEMATICAL PRACTICES
MP1, MP2, MP6

You can change an equation involving a fraction to an equation involving only whole numbers. To do so, multiply both sides of the equation by the denominator of the fraction.

Unlock the Problem

On canoe trips, people sometimes carry their canoes between bodies of water. Maps for canoeing use a unit of length called a *rod* to show distances. Victoria and Mick carry their canoe 40 rods. The equation $40 = \frac{2}{11}d$ can be used to find the distance d in yards that they carried the canoe. How many yards did they carry the canoe?

Use the graphic organizer to help you solve the problem.

Read the Problem	**Solve the Problem**
What do I need to find? I need to find _____ _____.	• Write a simpler equation. Write the equation. $\qquad 40 = \frac{2}{11}d$ Multiply both sides by $\qquad 11 \times 40 = $ _____ $\times \frac{2}{11}d$ the denominator. Multiply. _____ $= 2d$
What information do I need to use? I need to use _____.	• Solve the simpler equation. Write the equation. $\qquad 440 = 2d$
How will I use the information? I can solve a simpler problem by changing the equation to an equation involving only whole numbers. Then I can solve the simpler equation.	Use the Division Property of Equality. $\qquad \dfrac{440}{\ } = \dfrac{2d}{\ }$ Divide. _____ $= d$

So, Victoria and Mick carried their canoe _____ yards.

Math Talk

MATHEMATICAL PRACTICES ①

Evaluate How can you check that your answer to the problem is correct?

Chapter 8 457

If an equation contains more than one fraction, you can change it to an equation involving only whole numbers by multiplying both sides of the equation by the product of the denominators of the fractions.

🔓 Try Another Problem

Trevor is making $\frac{2}{3}$ of a recipe for chicken noodle soup. He adds $\frac{1}{2}$ cup of chopped celery. The equation $\frac{2}{3}c = \frac{1}{2}$ can be used to find the number of cups c of chopped celery in the original recipe. How many cups of chopped celery does the original recipe call for?

Use the graphic organizer to help you solve the problem.

Read the Problem	Solve the Problem
What do I need to find?	
What information do I need to use?	
How will I use the information?	

So, the original recipe calls for _____ cup of chopped celery.

- **MATHEMATICAL PRACTICE ⑥ Describe a Method** Describe another method that you could use to solve the problem.

_____

_____

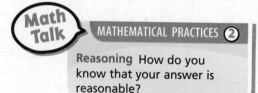

Math Talk

MATHEMATICAL PRACTICES ②

Reasoning How do you know that your answer is reasonable?

Name _____

Unlock the Problem
- ✓ Circle the important information.
- ✓ Use the Properties of Equality when you solve equations.
- ✓ Check your solution by substituting it into the original equation.

Share and Show

1. Connor ran 3 kilometers in a relay race. His distance represents $\frac{3}{10}$ of the total distance of the race. The equation $\frac{3}{10} d = 3$ can be used to find the total distance d of the race in kilometers. What was the total distance of the race?

   First, write a simpler equation by multiplying both sides by the denominator of the fraction.

   _____

   Next, solve the simpler equation.

   _____

   So, the race is _____ long.

2. **THINK SMARTER** **What if** Connor's distance of 3 kilometers represented only $\frac{2}{10}$ of the total distance of the race. What would the total distance of the race have been?

   _____

3. The lightest puppy in a litter weighs 9 ounces, which is $\frac{3}{4}$ of the weight of the heaviest puppy. The equation $\frac{3}{4} w = 9$ can be used to find the weight w in ounces of the heaviest puppy. How much does the heaviest puppy weigh?

   _____

4. Sophia took home $\frac{2}{5}$ of the pizza that was left over from a party. The amount she took represents $\frac{1}{2}$ of a whole pizza. The equation $\frac{2}{5} p = \frac{1}{2}$ can be used to find the number of pizzas p left over from the party. How many pizzas were left over?

   _____

5. A city received $\frac{3}{4}$ inch of rain on July 31. This represents $\frac{3}{10}$ of the total amount of rain the city received in July. The equation $\frac{3}{10} r = \frac{3}{4}$ can be used to find the amount of rain r in inches the city received in July. How much rain did the city receive in July?

   _____

WRITE ▶ *Math* • **Show Your Work**

© Houghton Mifflin Harcourt Publishing Company

On Your Own

6. **GO DEEPER** Carole ordered 4 dresses for $80 each, a $25 sweater, and a coat. The cost of the items without sales tax was $430. What was the cost of the coat?

_____

7. **THINK SMARTER** A dog sled race is 25 miles long. The equation $\frac{5}{8}k = 25$ can be used to estimate the race's length k in kilometers. Approximately how many hours will it take a dog sled team to finish the race if it travels at an average speed of 30 kilometers per hour?

_____

_____

8. **MATHEMATICAL PRACTICE 6** **Explain a Method** Explain how you could use the strategy *solve a simpler problem* to solve the equation $\frac{3}{4}x = \frac{3}{10}$.

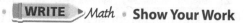
.... ║WRITE ▸ *Math* • **Show Your Work** ....

_____

_____

_____

_____

9. **THINK SMARTER** In a basket of fruit, $\frac{5}{6}$ of the pieces of fruit are apples. There are 20 apples in the display. The equation $\frac{5}{6}f = 20$ can be used to find how many pieces of fruit f are in the basket. Use words and numbers to explain how to solve the equation to find how many pieces of fruit are in the basket.

_____

_____

_____

_____

Problem Solving • Equations with Fractions

COMMON CORE STANDARD—6.EE.B.7
Reason about and solve one-variable equations and inequalities.

Read each problem and solve.

1. Stu is 4 feet tall. This height represents $\frac{6}{7}$ of his brother's height. The equation $\frac{6}{7}h = 4$ can be used to find the height h, in feet, of Stu's brother. How tall is Stu's brother?

$$7 \times \frac{6}{7}h = 7 \times 4$$

$$6h = 28$$

$$\frac{6h}{6} = \frac{28}{6}$$

$$h = 4\frac{2}{3}$$

$4\frac{2}{3}$ feet
_____

2. Bryce bought a bag of cashews. He served $\frac{7}{8}$ pound of cashews at a party. This amount represents $\frac{2}{3}$ of the entire bag. The equation $\frac{2}{3}n = \frac{7}{8}$ can be used to find the number of pounds n in a full bag. How many pounds of cashews were in the bag that Bryce bought?

_____

3. In Jaime's math class, 9 students chose soccer as their favorite sport. This amount represents $\frac{3}{8}$ of the entire class. The equation $\frac{3}{8}s = 9$ can be used to find the total number of students s in Jaime's class. How many students are in Jaime's math class?

_____

4. **WRITE** ▸*Math* Write a math problem for the equation. $\frac{3}{4}n = \frac{5}{6}$. Then solve a simpler problem to find the solution.

_____

_____

_____

Lesson Check (6.EE.B.7)

1. Roger served $\frac{5}{8}$ pound of crackers, which was $\frac{2}{3}$ of the entire box. What was the weight of the crackers originally in the box?

2. Bowser ate $4\frac{1}{2}$ pounds of dog food. That amount is $\frac{3}{4}$ of the entire bag of dog food. How many pounds of dog food were originally in the bag?

Spiral Review (6.RP.A.3d, 6.NS.A.1, 6.EE.A.2a, 6.EE.B.7)

3. What is the quotient $4\frac{2}{3} \div 4\frac{1}{5}$?

4. Miranda had 4 pounds, 6 ounces of clay. She divided it into 10 equal parts. How heavy was each part?

5. The amount Denise charges to repair computers is $50 an hour plus a $25 service fee. Write an expression to show how much she will charge for h hours of work.

6. Luis has saved $14 for a skateboard that costs $52. He can use the equation $14 + m = 52$ to find how much more money m he needs. How much more does he need?

FOR MORE PRACTICE
GO TO THE
Personal Math Trainer

✓ Mid-Chapter Checkpoint

Personal Math Trainer
Online Assessment and Intervention

Vocabulary

Choose the best term from the box to complete the sentence.

Vocabulary
equation
inverse operations
solution of an equation

1. A(n) _____ is a statement that two mathematical expressions are equal. (p. 421)

2. Adding 5 and subtracting 5 are _____. (p. 439)

Concepts and Skills

Write an equation for the word sentence. (6.EE.B.7)

3. The sum of a number and 4.5 is 8.2.

4. Three times the cost is $24.

_____

_____

Determine whether the given value of the variable is a solution of the equation. (6.EE.B.5)

5. $x - 24 = 58; x = 82$

6. $\frac{1}{3}c = \frac{3}{8}; c = \frac{3}{4}$

_____

_____

Solve the equation, and check the solution. (6.EE.B.7)

7. $a + 2.4 = 7.8$

8. $b - \frac{1}{4} = 3\frac{1}{2}$

_____

_____

9. $3x = 27$

10. $\frac{1}{3}s = \frac{1}{5}$

_____

_____

11. $\frac{t}{4} = 16$

12. $\frac{w}{7} = 0.3$

_____

_____

13. A stadium has a total of 18,000 seats. Of these, 7,500 are field seats, and the rest are grandstand seats. Write an equation that could be used to find the number of grandstand seats s. (6.EE.B.7)

_____

14. Aaron wants to buy a bicycle that costs \$128. So far, he has saved \$56. The equation $a + 56 = 128$ can be used to find the amount a in dollars that Aaron still needs to save. What is the solution of the equation? (6.EE.B.7)

_____

15. GO DEEPER Ms. McNeil buys 2.4 gallons of gasoline. The total cost is \$7.56. Write and solve an equation to find the price p in dollars of one gallon of gasoline. (6.EE.B.7)

_____

16. Crystal is picking blueberries. So far, she has filled $\frac{2}{3}$ of her basket, and the blueberries weigh $\frac{3}{4}$ pound. The equation $\frac{2}{3}w = \frac{3}{4}$ can be used to estimate the weight w in pounds of the blueberries when the basket is full. About how much will the blueberries in Crystal's basket weigh when it is full? (6.EE.B.7)

_____

Name _____

Solutions of Inequalities

Essential Question How do you determine whether a number is a solution of an inequality?

Common Core Expressions and Equations—
6.EE.B.5
MATHEMATICAL PRACTICES
MP2, MP3, MP6

An **inequality** is a mathematical sentence that compares two expressions using the symbol $<$, $>$, $\leq$, $\geq$, or $\neq$. These are examples of inequalities:

$$8 < 11 \qquad 9 > {}^-4 \qquad a \leq 50 \qquad x \geq 3.2$$

A **solution of an inequality** is a value of a variable that makes the inequality true. Inequalities can have more than one solution.

> **Math Idea**
> - The symbol $\leq$ means "is less than or equal to."
> - The symbol $\geq$ means "is greater than or equal to."

🔑 Unlock the Problem · Real World

A library has books from the Middle Ages. The books are more than 650 years old. The inequality $a > 650$ represents the possible ages a in years of the books. Determine whether $a = 678$ or $a = 634$ is a solution of the inequality, and tell what the solution means.

Use substitution to determine the solution.

STEP 1 Check whether $a = 678$ is a solution.

Write the inequality. $\qquad\qquad a > 650$

Substitute 678 for a. $\qquad\qquad \underline{\hspace{2cm}} \overset{?}{>} 650$

Compare the values. $\qquad$ 678 is $\underline{\hspace{2cm}}$ than 650.

The inequality is true when $a = 678$, so $a = 678$ is a solution.

STEP 2 Check whether $a = 634$ is a solution.

Write the inequality. $\qquad\qquad a > 650$

Substitute 634 for a. $\qquad\qquad \underline{\hspace{2cm}} \overset{?}{>} 650$

Compare the values. $\qquad$ 634 $\underline{\hspace{2cm}}$ greater than 650.

The inequality $\underline{\hspace{2cm}}$ true when $a = 634$, so $a = 634$ $\underline{\hspace{2cm}}$ a solution.

The solution $a = 678$ means that a book in the library from the

Middle Ages could be $\underline{\hspace{2cm}}$ years old.

Math Talk MATHEMATICAL PRACTICES ③

Apply Give another solution of the inequality $a > 650$. Explain how you determined the solution.

🔑 Example 1 Determine whether the given value of the variable is a solution of the inequality.

Ⓐ $b < 0.3; b = {}^-0.2$

Write the inequality. $b < 0.3$

Substitute the given value for the variable.   _____ $\overset{?}{<}$ 0.3

Compare the values. $^-0.2$ is _____ than 0.3.

The inequality _____ true when $b = {}^-0.2$, so $b = {}^-0.2$ _____
a solution.

Ⓑ $m \geq \frac{2}{3}; m = \frac{3}{5}$

Write the inequality. $m \geq \frac{2}{3}$

Substitute the given value for
the variable.   _____ $\overset{?}{\geq} \frac{2}{3}$

Rewrite the fractions with a
common denominator. $\overset{?}{\geq}$ ——
                                                                                $\overline{15}$ $\overline{15}$

Compare the values. $\frac{9}{15}$ _____ greater than or equal to $\frac{10}{15}$.

The inequality _____ true when $m = \frac{3}{5}$, so $m = \frac{3}{5}$ _____
a solution.

🔑 Example 2

An airplane can hold no more than 416 passengers. The inequality $p \leq 416$ represents the possible number of passengers p on the airplane, where p is a whole number. Give two solutions of the inequality, and tell what the solutions mean.

Think: The solutions of the inequality are whole numbers _____ than or

_____ to 416.

• $p = 200$ is a solution because 200 is _____ than _____.

• $p = $ _____ is a solution because _____ is _____
than 416.

These solutions mean that the number of passengers on the

plane could be _____ or _____.

© Houghton Mifflin Harcourt Publishing Company

MATHEMATICAL PRACTICES ③

Apply Give an example of a value of p that is not a solution of the inequality. How do you know it is not a solution?

Name _____

Determine whether the given value of the variable is a solution of the inequality.

1. $a \geq {}^-6; a = {}^-3$

_____ $\overset{?}{\geq}$ $^-6$

2. $y < 7.8; y = 8$

3. $c > \frac{1}{4}; c = \frac{1}{5}$

_____

_____

4. $x \leq 3; x = 3$

5. $d < {}^-0.52; d = {}^-0.51$

6. $t \geq \frac{2}{3}; t = \frac{3}{4}$

_____

_____

_____

Math Talk

MATHEMATICAL PRACTICES ⑥

Explain How could you use a number line to check your answer to Exercise 5?

On Your Own

Practice: Copy and Solve Determine whether $s = \frac{3}{5}$, $s = 0$, or $s = 1.75$ are solutions of the inequality.

7. $s > {}^-1$

8. $s \leq 1\frac{2}{3}$

9. $s < 0.43$

Give two solutions of the inequality.

10. $e < 3$

11. $p > {}^-12$

12. $y \geq 5.8$

13. MATHEMATICAL PRACTICE ② **Connect Symbols and Words** A person must be at least 18 years old to vote. The inequality $a \geq 18$ represents the possible ages a in years at which a person can vote. Determine whether $a = 18$, $a = 17\frac{1}{2}$, and $a = 91.5$ are solutions of the inequality, and tell what the solutions mean.

_____

_____

Problem Solving • Applications

The table shows ticket and popcorn prices at five movie theater chains. Use the table for 14–15.

14. **GO DEEPER** The inequality $p < 4.75$ represents the prices p in dollars that Paige is willing to pay for popcorn. The inequality $p < 8.00$ represents the prices p in dollars that Paige is willing to pay for a movie ticket. At how many theaters would Paige be willing to buy a ticket and popcorn?

_____

15. **THINK SMARTER** **Sense or Nonsense?** Edward says that the inequality $d \geq 4.00$ represents the popcorn prices in the table, where d is the price of popcorn in dollars. Is Edward's statement sense or nonsense? Explain.

_____

_____

Movie Theater Prices

Ticket Price ($)	Popcorn Price ($)
8.00	4.25
8.50	5.00
9.00	4.00
7.50	4.75
7.25	4.50

16. **MATHEMATICAL PRACTICE ⑥** **Use Math Vocabulary** Explain why the statement $t > 13$ is an inequality.

_____

_____

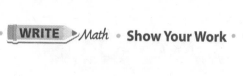

WRITE Math • **Show Your Work**

17. **THINK SMARTER +** The minimum wind speed for a storm to be considered a hurricane is 74 miles per hour. The inequality $w \geq 74$ represents the possible wind speeds of a hurricane.

Two possible solutions for the inequality $w \geq 74$

are | 71 / 73 / 75 | and | 80. / 60. / 40. |

© Houghton Mifflin Harcourt Publishing Company • Image Credits: (t) ©Fuse/Getty Images

Solutions of Inequalities

Common Core **COMMON CORE STANDARD—6.EE.B.5**
Reason about and solve one-variable equations and inequalities.

Determine whether the given value of the variable is a solution of the inequality.

1. $s \geq {}^-1; s = 1$

$$1 \overset{?}{\geq} {}^-1$$

solution

_____

2. $p < 0; p = 4$

_____

3. $y \leq {}^-3; y = {}^-1$

_____

4. $u > -\frac{1}{2}; u = 0$

_____

5. $q \geq 0.6; q = 0.23$

_____

6. $b < 2\frac{3}{4}; b = \frac{2}{3}$

_____

Give two solutions of the inequality.

7. $k < 2$

_____

8. $z \geq {}^-3$

_____

9. $f \leq {}^-5$

_____

Problem Solving (Real World)

10. The inequality $s \geq 92$ represents the score s that Jared must earn on his next test to get an A on his report card. Give two possible scores that Jared could earn to get the A.

_____

11. The inequality $m \leq \$20$ represents the amount of money that Sheila is allowed to spend on a new hat. Give two possible money amounts that Sheila could spend on the hat.

_____

12. **WRITE** ▸*Math* Describe a situation and write an inequality to represent the situation. Give a number that is a solution and another number that is not a solution of the inequality.

_____

_____

Lesson Check

1. Three of the following are solutions of $g < {}^-1\frac{1}{2}$. Which one is not a solution?

$g = {}^-4$ $g = {}^-7\frac{1}{2}$ $g = 0$ $g = {}^-2\frac{1}{2}$

_____

_____

2. The inequality $w \geq 3.2$ represents the weight of each pumpkin, in pounds, that is allowed to be picked to be sold. The weights of pumpkins are listed. How many pumpkins can be sold? Which pumpkins can be sold?

3.18 lb, 4 lb, 3.2 lb, 3.4 lb, 3.15 lb

_____

_____

Spiral Review

3. What is the value of $8 + (27 \div 9)^2$?

_____

4. Write an expression that is equivalent to $5(3x + 2z)$.

_____

5. Tina bought a t-shirt and sandals. The total cost was $41.50. The t-shirt cost $8.95. The equation $8.95 + c = 41.50$ can be used to find the cost c in dollars of the sandals. How much did the sandals cost?

_____

6. Two-thirds of a number is equal to 20. What is the number?

_____

FOR MORE PRACTICE GO TO THE
Personal Math Trainer

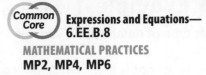

Write Inequalities

Essential Question How do you write an inequality to represent a situation?

 Expressions and Equations—6.EE.B.8
MATHEMATICAL PRACTICES
MP2, MP4, MP6

CONNECT You can use what you know about writing equations to help you write inequalities.

 Unlock the Problem Real World

The highest temperature ever recorded at the South Pole was 8°F. Write an inequality to show that the temperature *t* in degrees Fahrenheit at the South Pole is less than or equal to 8°F.

🔑 **Write an inequality for the situation.**

• Underline the words that tell you which inequality symbol to use.

• Will you use an equal sign in your inequality? Explain.

_____

_____

Think:

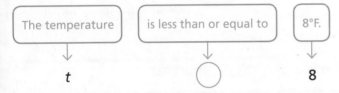

The temperature	is less than or equal to	8°F.
↓	↓	↓
t	◯	8

So, an inequality that describes the temperature *t* in

degrees Fahrenheit at the South Pole is _____ .

Try This! The directors of an animal shelter need to raise more than **$50,000** during a fundraiser. Write an inequality that represents the amount of money *m* in dollars that the directors need to raise.

Think:

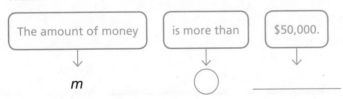

The amount of money	is more than	$50,000.
↓	↓	↓
m	◯	_____

So, an inequality that describes the amount of money *m* in

dollars is _____ .

 Math Talk

MATHEMATICAL PRACTICES ⑥

Explain How did you know which inequality symbol to use in the Try This! problem?

🔑 Example 1 Write an inequality for the word sentence. Tell what type of numbers the variable in the inequality can represent.

Ⓐ The weight is less than $3\frac{1}{2}$ pounds.

Think: Let w represent the unknown weight in pounds.

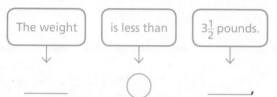

| The weight | is less than | $3\frac{1}{2}$ pounds. |

_____ ◯ _____ , where w is a positive number

Ⓑ There must be at least 65 police officers on duty.

Think: Let p represent the number of police officers. The phrase "at least" is

equivalent to "is _____ than or equal to."

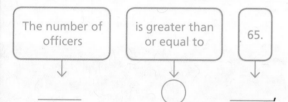

| The number of officers | is greater than or equal to | 65. |

_____ ◯ _____ , where p is a _____ number

Math Talk

MATHEMATICAL PRACTICES ②

Reasoning Explain why the value of p must be a whole number.

🔑 Example 2 Write two word sentences for the inequality.

Ⓐ $n \leq 0.3$

- n is _____ than or _____ to 0.3.

- n is no _____ than 0.3.

Ⓑ $a > {}^-4$

- a is _____ than $^-4$.

- a is _____ than $^-4$.

- **THINK SMARTER** Which inequality symbol would you use to show that the number of people attending a party will be at most 14? Explain.

_____

_____

472

Name _____

Write an inequality for the word sentence. Tell what type of numbers the variable in the inequality can represent.

1. The elevation e is greater than or equal to 15 meters.

_____

2. A passenger's age a must be more than 4 years.

_____

Write a word sentence for the inequality.

3. $b < \frac{1}{2}$

_____

4. $m \geq 55$

_____

On Your Own

5. MATHEMATICAL PRACTICE ⑥ **Compare** Explain the difference between $t \leq 4$ and $t < 4$.

_____

_____

6. GO DEEPER A children's roller coaster is limited to riders whose height is at least 30 inches and at most 48 inches. Write two inequalities that represent the height h of riders for the roller coaster.

_____

_____

7. THINK SMARTER Match the inequality with the word sentence it represents.

$r > 10$ •   • Walter sold more than 10 tickets.

$s \leq 10$ •   • Fewer than 10 children are at the party.

$t \geq 10$ •   • No more than 10 people can be seated at a table.

$w < 10$ •   • At least 10 people need to sign up for the class.

Connect to Reading

Make Generalizations

The reading skill *make generalizations* can help you write inequalities to represent situations. A generalization is a statement that is true about a group of facts.

Sea otters spend almost their entire lives in the ocean. Their thick fur helps them to stay warm in cold water. Sea otters often float together in groups called *rafts*. A team of biologists weighed the female sea otters in one raft off the coast of Alaska. The chart shows their results.

Write two inequalities that represent generalizations about the sea otter weights.

First, list the weights in pounds in order from least to greatest.

50, 51, 54, _____, _____, _____, _____, _____,

_____, _____, _____, _____

Next, write an inequality to describe the weights by using the least weight in the list. Let *w* represent the weights of the otters in pounds.

Think: The least weight is _____ pounds, so all of the weights are greater than or equal to 50 pounds.

$w \bigcirc 50$

Now, write an inequality to describe the weights by using the greatest weight in the list.

Think: The greatest weight is _____ pounds, so

$w \bigcirc 71$

all of the weights are _____ than or equal to

_____ pounds.

So, the inequalities _____ and _____ represent generalizations about the weights *w* in pounds of the otters.

8. **THINK SMARTER** Use the chart at the right to write two inequalities that represent generalizations about the number of sea otter pups per raft.

_____

_____

Weights of Female Sea Otters	
Otter Number	Weight (pounds)
1	50
2	61
3	62
4	69
5	71
6	54
7	68
8	62
9	58
10	51
11	61
12	66

Sea Otter Pups per Raft	
Raft Number	Number of Pups
1	7
2	10
3	15
4	23
5	6
6	16
7	20
8	6

Write Inequalities

Common Core

COMMON CORE STANDARD—6.EE.B.8
*Reason about and solve one-variable equations
and inequalities.*

**Write an inequality for the word sentence. Tell what type of
numbers the variable in the inequality can represent.**

1. The width w is greater than 4 centimeters.

The inequality symbol for "is greater than" is $>$. $w > 4$, where w is the width in

centimeters. w is a positive number.

2. The score s in a basketball game is greater than
or equal to 10 points.

3. The mass m is less than 5 kilograms.

4. The height h is greater than 2.5 meters.

5. The temperature t is less than or equal to $^-3°$.

Write a word sentence for the inequality.

6. $k < {}^-7$

7. $z \geq 2\frac{3}{5}$

Problem Solving · Real World

8. Tabby's mom says that she must read for at
least 30 minutes each night. If m represents the
number of minutes reading, what inequality can
represent this situation?

9. Phillip has a $25 gift card to his favorite
restaurant. He wants to use the gift card to
buy lunch. If c represents the cost of his lunch,
what inequality can describe all of the possible
amounts of money, in dollars, that Phillip can
spend on lunch?

10. **WRITE** ▸*Math* Write a short paragraph explaining to a
new student how to write an inequality.

Lesson Check (6.EE.B.8)

1. At the end of the first round in a quiz show, Jeremy has at most ⁻20 points. Write an inequality that means "at most ⁻20".

_____

_____

_____

2. Describe the meaning of $y \geq 7.9$ in words.

_____

_____

_____

Spiral Review (6.EE.A.2a, 6.EE.A.4, 6.EE.B.5, 6.EE.B.7)

3. Let y represent Jaron's age in years. If Dawn were 5 years older, she would be Jaron's age. Which expression represents Dawn's age?

_____

4. Simplify the expression $7 \times 3g$.

_____

5. What is the solution of the equation $8 = 8f$?

_____

6. Which of the following are solutions of the inequality $k \leq {}^{-}2$?

$k = 0, k = {}^{-}2, k = {}^{-}4, k = 1, k = {}^{-}1\frac{1}{2}$

_____

© Houghton Mifflin Harcourt Publishing Company

FOR MORE PRACTICE
GO TO THE
Personal Math Trainer

Graph Inequalities

Essential Question How do you represent the solutions of an inequality on a number line?

 Common Core **Expressions and Equations—6.EE.B.8**

MATHEMATICAL PRACTICES
MP3, MP4, MP6

Inequalities can have an infinite number of solutions. The solutions of the inequality $x > 2$, for example, include all numbers greater than 2. You can use a number line to represent all of the solutions of an inequality.

The number line at right shows the solutions of the inequality $x > 2$.

$x > 2$

The empty circle at 2 shows that 2 is not a solution. The shading to the right of 2 shows that values greater than 2 are solutions.

 Unlock the Problem (Real World)

Forest fires are most likely to occur when the air temperature is greater than 60°F. The inequality $t > 60$ represents the temperatures t in degrees Fahrenheit for which forest fires are most likely. Graph the solutions of the inequality on a number line.

🔑 **Show the solutions of $t > 60$ on a number line.**

Think: I need to show all solutions that are greater than 60.

Draw an empty circle at _____ to show that 60 is not a solution.

Shade to the _____ of _____ to show that values greater than 60 are solutions.

```
  +--+--+--+--+--+--+--+--+--+--+
  0  10 20 30 40 50 60 70 80 90 100
```

Try This! Graph the solutions of the inequality $y < 5$.

Draw an empty circle at _____ to show that 5 is not a solution.

Shade to the _____ of _____ to show that values less than 5 are solutions.

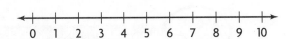

```
  +--+--+--+--+--+--+--+--+--+--+
  0  1  2  3  4  5  6  7  8  9  10
```

- **MATHEMATICAL PRACTICE 6** **Make Connections** Explain why $y = 5$ is not a solution of the inequality $y < 5$.

You can also use a number line to show the solutions of an inequality that includes the symbol ≤ or ≥.

The number line at right shows the solutions of the inequality $x \geq 2$.

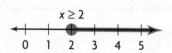

The filled-in circle at 2 shows that 2 is a solution. The shading to the right of 2 shows that values greater than 2 are also solutions.

Example 1 Graph the solutions of the inequality on a number line.

A $w \leq 0.8$

Draw a filled-in circle at _____ to show that 0.8 is a solution.

Shade to the _____ of _____ to show that values less than 0.8 are also solutions.

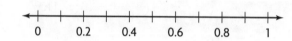

B $n \geq {}^-3$

Draw a filled-in circle at _____ to show that $^-3$ is a solution.

Shade to the _____ of _____ to show that values greater than $^-3$ are also solutions.

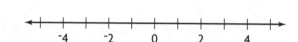

Example 2 Write the inequality represented by the graph.

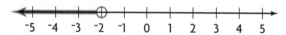

Use x (or another letter) for the variable in the inequality.

The _____ circle at _____ shows that $^-2$

_____ a solution.

The shading to the _____ of _____ shows that values

_____ than $^-2$ are solutions.

So, the inequality represented by the graph is _____.

Math Talk

MATHEMATICAL PRACTICES ⑥

Explain How do you know whether to shade to the right or to the left when graphing an inequality?

478

Name _____

Graph the inequality.

1. $m < 15$

Draw an empty circle at _____ to show that 15 is

not a solution. Shade to the _____ of _____ to
show that values less than 15 are solutions.

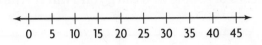

2. $c \geq {}^{-}1.5$

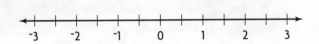

3. $b \leq \dfrac{5}{8}$

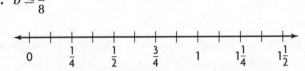

On Your Own

Math Talk | MATHEMATICAL PRACTICES ❸

Apply Why is it easier to graph the solutions of an inequality than it is to list them?

Practice: Copy and Solve Graph the inequality.

4. $a < \dfrac{2}{3}$

5. $x > {}^{-}4$

6. $k \geq 0.3$

7. $t \leq 6$

Write the inequality represented by the graph.

8.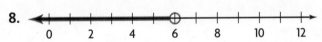

_____

9.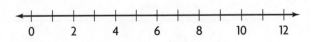

_____

10. MATHEMATICAL PRACTICE ④ **Model Mathematics** The inequality $w \geq 60$ represents the wind speed w in miles per hour of a tornado. Graph the solutions of the inequality on the number line.

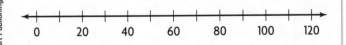

11. GO DEEPER Graph the solutions of the inequality $c < 12 \div 3$ on the number line.

Problem Solving • Applications (Real World)

The table shows the height requirements for rides at an amusement park. Use the table for 12–16.

12. Write an inequality representing *t*, the heights in inches of people who can go on Twirl & Whirl.

_____

13. Graph your inequality from Exercise 12.

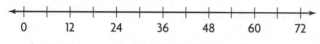

14. Write an inequality representing *r*, the heights in inches of people who can go on Race Track.

_____

15. Graph your inequality from Exercise 14.

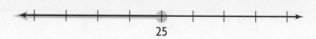

Height Requirements	
Ride	Minimum height (in.)
Mighty Mountain	44
Race Track	42
River Rapids	38
Twirl & Whirl	48

16. **THINK SMARTER** Write an inequality representing *b*, the heights in inches of people who can go on *both* River Rapids and Mighty Mountain. Explain how you determined your answer.

WRITE ▸ Math • Show Your Work

_____

_____

_____

17. **THINK SMARTER** Alena graphed the inequality $c \leq 25$.

Darius said that 25 is not part of the solution of the inequality. Do you agree or disagree with Darius? Use numbers and words to support your answer.

25

_____

_____

Graph Inequalities

COMMON CORE STANDARD—6.EE.B.8
Reason about and solve one-variable equations and inequalities.

Graph the inequality.

1. $h \geq 3$

Draw a filled-in circle at ___3___ to show that 3 is a solution. Shade to the ___right___ of ___3___ to show that values greater than 3 are solutions.

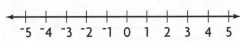

2. $x < \frac{-4}{5}$

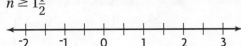

3. $y > {}^-2$

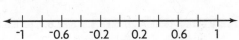

4. $n \geq 1\frac{1}{2}$

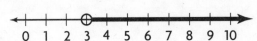

5. $c \leq {}^-0.4$

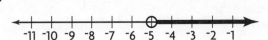

Write the inequality represented by the graph.

6.

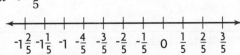

_____

7.

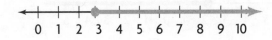

_____

 Problem Solving

8. The inequality $x \leq 2$ represents the elevation x of a certain object found at a dig site. Graph the solutions of the inequality on the number line.

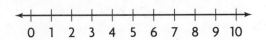

9. The inequality $x \geq 144$ represents the possible scores x needed to pass a certain test. Graph the solutions of the inequality on the number line.

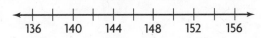

10. **WRITE** *Math* Write an inequality and graph the solutions on a number line.

_____

_____

Lesson Check (6.EE.B.8)

1. Write the inequality that is shown by the graph.

2. Describe the graph of $g < 0.6$.

Spiral Review (6.EE.A.2b, 6.EE.B.5, 6.EE.B.7, 6.EE.B.8)

3. Write an expression that shows the product of 5 and the difference of 12 and 9.

4. What is the solution of the equation $8.7 + n = 15.1$?

5. The equation $12x = 96$ gives the number of egg cartons x needed to package 96 eggs. Solve the equation to find the number of cartons needed.

6. The lowest price on an MP3 song is $0.35. Write an inequality that represents the cost c of an MP3 song.

FOR MORE PRACTICE
GO TO THE
Personal Math Trainer

✓ Chapter 8 Review/Test

Personal Math Trainer
Online Assessment
and Intervention

1. For numbers 1a–1c, choose Yes or No to indicate whether the given value of the variable is a solution of the equation.

    1a. $\frac{2}{5}v = 10$; $v = 25$ ○ Yes   ○ No

    1b. $n + 5 = 15$; $n = 5$ ○ Yes   ○ No

    1c. $5z = 25$; $z = 5$ ○ Yes   ○ No

2. The distance from third base to home plate is 88.9 feet. Romeo was 22.1 feet away from third base when he was tagged out. The equation $88.9 - t = 22.1$ can be used to determine how far he needed to run to get to home plate. Using substitution, the coach determines that Romeo needed

to run
66
66.8
111
feet to get to home plate.

3. There are 84 grapes in a bag. Four friends are sharing the grapes. Write an equation that can be used to find out how many grapes g each friend will get if each friend gets the same number of grapes.

_____

4. Match each scenario with the equation that can be used to solve it.

| Jane's dog eats 3 pounds of food a week. How many weeks will a 24-pound bag last? | • | |
| | | • $3x = 39$ |

| There are 39 students in the gym, and there are an equal number of students in each class. If three classes are in the gym, how many students are in each class? | • | |
| | | • $4x = 24$ |

| There are 4 games at the carnival. Kevin played all the games in 24 minutes. How many minutes did he spend at each game if he spent an equal amount of time at each? | • | |
| | | • $3x = 24$ |

GO DIGITAL Assessment Options
Chapter Test

5. **GO DEEPER** Frank's hockey team attempted 15 more goals than Spencer's team. Frank's team attempted 23 goals. Write and solve an equation that can be used to find how many goals Spencer's team attempted.

_____

6. Ryan solved the equation $10 + y = 17$ by drawing a model. Use numbers and words to explain how Ryan's model can be used to find the solution.

7. Gabriella and Max worked on their math project for a total of 6 hours. Max worked on the project for 2 hours by himself. Solve the equation $x + 2 = 6$ to find out how many hours Gabriella worked on the project.

_____

8. Select the equations that have the solution $m = 17$. Mark all that apply.

(A) $3 + m = 21$

(B) $m - 2 = 15$

(C) $14 = m - 3$

(D) $2 = m - 15$

9. Describe how you could use algebra tiles to model the equation $4x = 20$.

10. For numbers 10a–10d, choose Yes or No to indicate whether the equation has the solution $x = 12$.

10a. $\frac{3}{4}x = 9$ ○ Yes ○ No

10b. $3x = 36$ ○ Yes ○ No

10c. $5x = 70$ ○ Yes ○ No

10d. $\frac{x}{3} = 4$ ○ Yes ○ No

11. Bryan rides the bus to and from work on the days he works at the library. In one month, he rode the bus 24 times. Solve the equation $2x = 24$ to find the number of days Bryan worked at the library. Use a model.

12. Betty needs $\frac{3}{4}$ of a yard of fabric to make a skirt. She bought 9 yards of fabric.

Part A

Write and solve an equation to find how many skirts x she can make from 9 yards of fabric.

Part B

Explain how you determined which operation was needed to write the equation.

13. Karen is working on her math homework. She solves the equation $\frac{b}{8} = 56$ and says that the solution is $b = 7$. Do you agree or disagree with Karen? Use words and numbers to support your answer. If her answer is incorrect, find the correct answer.

14. There are 70 historical fiction books in the school library. Historical fiction books make up $\frac{1}{10}$ of the library's collection. The equation $\frac{1}{10}b = 70$ can be used to find out how many books the library has. Solve the equation to find the total number of books in the library's collection. Use numbers and words to explain how to solve $\frac{1}{10}b = 70$.

15. Andy drove 33 miles on Monday morning. This was $\frac{3}{7}$ of the total number of miles he drove on Monday. Solve the equation $\frac{3}{7}m = 33$ to find the total number of miles Andy drove on Monday.

Personal Math Trainer

16. **THINK SMARTER +** The maximum number of players allowed on a lacrosse team is 23. The inequality $t \leq 23$ represents the total number of players t allowed on the team.

Two possible solutions for the inequality are
| 23 |
| 25 | and
| 27 |

| 26. |
| 24. |
| 22. |

17. Mr. Charles needs to have at least 10 students sign up for homework help in order to use the computer lab. The inequality $h \geq 10$ represents the number of students h who must sign up. Select possible solutions of the inequality. Mark all that apply.

Ⓐ 7

Ⓑ 8

Ⓒ 9

Ⓓ 10

Ⓔ 11

Ⓕ 12

18. The maximum capacity of the school auditorium is 420 people. Write an inequality for the situation. Tell what type of numbers the variable in the inequality can represent.

_____

19. Match the inequality to the word sentence it represents.

$w < 70$ •

$x \leq 70$ •

$y > 70$ •

$z \geq 70$ •

• | The temperature did not drop below 70 degrees. |

• | Dane saved more than $70. |

• | Fewer than 70 people attended the game. |

• | No more than 70 people can participate. |

20. Cydney graphed the inequality $d \leq 14$.

14

Part A

Dylan said that 14 is not a solution of the inequality. Do you agree or disagree with Dylan? Use numbers and words to support your answer.

Part B

Suppose Cydney's graph had an empty circle at 14. Write the inequality represented by this graph.

_____